HOW FLYING BEES PILOT AND OTHER ARTHROPOD WONDERS
Adrian Horridge

ISBN: 978-1-914934-42-1

Text and graphics: Adrian Horridge. www.adrian-horridge.org.

Published by Northern Bee Books 2022
Northern Bee Books, Scout Bottom Farm, Mytholmroyd,
Hebden Bridge HX7 5JS (UK).
www.northernbeebooks.co.uk
+44 (0) 1422 882751.

Book design by www.SiPat.co.uk

HOW FLYING BEES PILOT
AND OTHER ARTHROPOD WONDERS

Adrian Horridge

HOW FLYING BEES PILOT AND OTHER ARTHROPOD WONDERS

Contents

What's it all about?

In the visual system of the honeybee, feature detectors for the amount of blue and distribution of green vertical edges were slowly revealed over the past century, as outlined in a companion booklet (Horridge, 2021). The method was to train bees on one of many targets and then test the trained bees until they failed when the last remaining cue or feature was removed. This was proof of a corresponding feature detector in the honeybee visual system. However, this told us nothing about how they detect the structure of surrounding space and how to pilot themselves without training.

Over the past century, a variety of innate behaviour and responses of the honeybee have been well described, some of which have no obvious function, others are remarkably sensitive. When they are consolidated together, it becomes obvious that honeybees, and presumably all insects, detect relative positions of the nearest objects by parallax when the eyes move, just like an astronomer plots star distances from changes in relative position as the earth moves. This gives honeybees abundant information for piloting themselves in flight.

Summary

1. Bees, and many other insects, have several easily reproducible behaviour patterns that are related to the avoidance of obstacles while flying. All insects experience continual flicker as edges cross the retina as they scan in flight, but they are not able to distinguish different flicker frequencies when the product of the light flux times the area is the same. They take a sum in each scan and it is simply a matter of total power in the flicker.

2. When all the surrounding panorama is rotated together around them, as when in a drum, most animals, including bees, follow the motion down to extremely low speeds. The use of this optomotor response has been a puzzle for a century.

3. Bees detect the shift in flicker power caused by any change in positions of edges, not the angular speed of the motion.

4. When the panorama, or even a single bar on a clean background, is moved around them during a short dark period of a few minutes, they return to their original position relative to it when the light returns. Therefore, they have a short-term retinotopic memory of the positions of vertical edges.

5. Even at low light levels, this allows them to respond to slow motion down to 15°/hour, which is the angular speed of the earth, relative to the sky. This provides a compass for some nocturnal insects and crabs, and they detect the continual slow motions of edges of shadows.

6. Flying bees avoid a rotating pattern that generates motion of edges and flicker up to 200Hz, and will not land on opening parallax (edges separating), which resembles the expanding or looming before a crash. They are attracted spontaneously to closing parallax (edges approaching and overlapping) where they willingly land. This shows that they have feature detectors for closing parallax.

7. We now see that the optomotor response holds still the relative motion of the distant panorama so that parallax can give a measure of the ranges of surrounding objects.

8. We failed to incorporate these principles into applications for flying drones and other vehicles because it is difficult to correlate the successive frames generated too rapidly in a digital camera.

1. A variety of gymnastics

How often have we watched as our bees emerge from their hive and after a short pause, rise and head off in one direction to their foraging place for the day. We marvel at their skill in dodging twigs and landing on flower after flower, despite being buffeted by gusts of wind. If we take a few bees and mark them with quick-dry paint, then release them, they circle a few times and come back directly to their hive. These are not mysterious feats. With much more, in recent years their control of flight has been analysed in great detail, and the results are applicable to unmanned vehicles with a computer on board.

For an insect, piloting implies stability in flight, navigation by signs in the sky and landmarks, measure range to obstacles and avoid them, control flight height and flight speed, land, return home; learn a measure of the distance of the homeward trip, and in honeybees convey some information to other bees. All these related topics are controlled by vision, but here I exclude foraging and feeding behaviour, which has been discussed in a recent publication (Horridge, 2021).

2. Which way is up, and where is home?

For a flying bee, the ultraviolet light (UV) of the sky shows which way is up, as demonstrated by bees that make a forward somersault when they fly over a shiny metal mirror laid flat. The UV originates from sunlight scattered by small particles in the air. Elongated particles float horizontally, so this scattered light is partly polarised in the plane that contains the scattered ray and the ray from the sun. The pattern in the blue sky formed by polarization indicates the position of the sun (Figure 1D).

Honeybees, crickets and some other insects that have a home territory, detect this pattern of polarization with a specialized row of UV-sensitive ommatidia along the top edge of each eye (Figure 1A, B), to locate the sun even behind a cloud. Bees learn this sky pattern when they first become foragers, and they associate it with the direction towards home and each obvious landmark in their familiar area, but there is no firm evidence that UV is remembered as a colour. Many insects, like dragonflies, damselflies, mantids, and large butterflies, also defend a home territory that they patrol, but are not investigated in detail.

Most insects will not fly in absolute darkness; a bee in a large illuminated white bag, with no contrasts visible, only walks. At the top of the head, bees and many other insects have a tiny cluster of three small eyes, called ocelli, that certainly assist locusts and dragonflies (the only ones studied) to detect the horizon, and with these they stabilise flight at night by starlight alone.

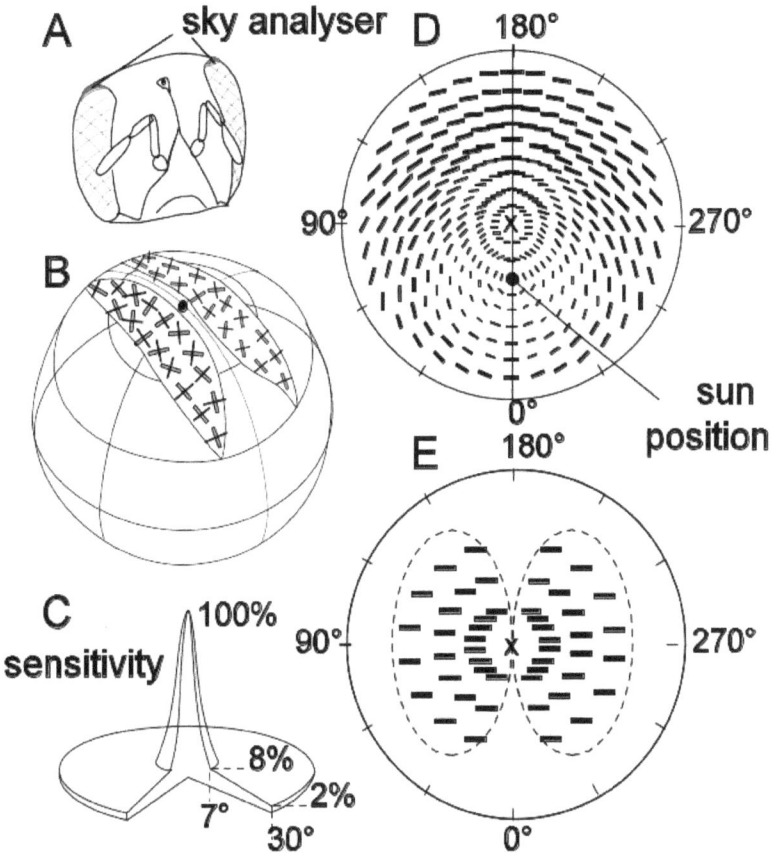

Figure 1. The sky compass is a pattern of ultraviolet polarization. A. Location of ommatidia at the dorsal edge of the eye. B. Orientations of pairs of specialized receptors in the special ommatidia. C. Broad angular sensitivity of a special receptor to the polarisation plane. D. The orientation pattern in the UV of the sky. E. Orientations of the central detectors, inferred from behaviour. Redrawn from Labhart (1980).

3. The response to flicker. Talbot's Law

A disturbed honeybee always tries to escape towards the brightest light, an innate action known as phototaxis, sometimes used mistakenly to test for colour vision. The attraction of a light source increases in proportion to its brightness. However, the structure of apposition eyes, with each ommatidium at a small angle to its neighbours, means that all visual input is continually flickered, as the head and eyes continually move relative to the panorama, so flicker cannot be avoided, and yet a small change in position against the background is immediately detected.

In 1935, Ernst Wolf and his wife Gertrude Zerrahn, who had both escaped from Europe to the USA, showed that above a critical flicker frequency of about 66Hz (flicks per second), bees could not distinguish between two windows that were steady, or flickering at a constant rate, when the product of average light emission multiplied by area was equal. That is, when each source generated the same average flux whatever the frequency. This is Talbot's Law, which applies to all visual systems investigated, supporting the finding that the bee and other primitive visual systems measure average intensity and area (total light), irrespective of temporal or spatial pattern.

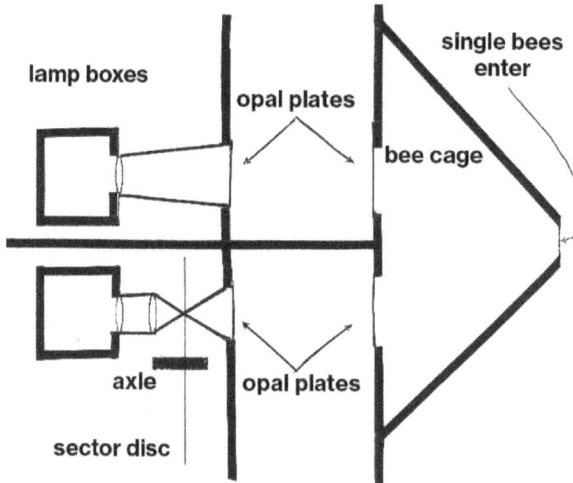

Figure 2. The apparatus used to test phototaxis (attraction to light). Individual bees selected one of the two windows. The optics excludes any motion stimulus, leaving only flicker. In their immediately preceding paper, both lights were flickering, with a similar result. (Wolf and Zerrahn-Wolf, 1935).

Clearly, honeybees are not frightened by pure flicker. In reports where bees distinguish flicker frequency, the calibration of the lights must be wrong, and the authors have not read the literature. This result also implies that the bee has a constant and reliable detection mechanism that measures and could learn the sum of the modulation detected in each visual scan in flight; both important points for our future conclusions.

4. The optomotor response

Insects that hover in the air, float on water, or stand still, are stabilized in position by the contrasting edges around them, as shown by the way they turn to follow when everything around them is rotated together. A large cardboard drum can be mounted on a bicycle wheel with a stationary platform on the top of the vertical axle at the centre (Figure 3B). An insect, for example, can be fixed with its head at the centre. The drum is lined on the inside with paper with equal black and white vertical stripes. Alternative papers for lining have different repeat periods of the stripes. A light straw can be glued to the insect's head to magnify rotations of the head on the neck. In this situation, a bee, grasshopper, blowfly, butterfly and some beetles, or even a cat, dog, or a human, consistently follows the drum's motion with eyes or head, in what has long been known as the optomotor response. The angular resolution of bars is easily measured by varying the stripe period.

In the early years of the 20th century, in dozens of investigations, the optomotor response was thought to be the principal reflex that enabled flying insects to keep on a straight course, or stationary when hovering. However, there were always questions as to how the animal could turn voluntarily when the eyes were locked on to the general surroundings, and how the head or whole insect could return to its former position in space after being moved. There were no answers, but many other relevant results were obtained.

First, the optomotor response is too slow to respond to fast displacements. Peak response in the fly was near 2Hz (pairs of bars passing any point per second), and cut off near 20Hz, but was only 3Hz in roll and lift. Therefore, most naturally imposed motions will have no effect. In fact, house flies turn in flight by making a sudden turn (called a saccade) that is too fast for the optomotor system to respond; then they continue on a new stabilised straight path. The optomotor response in the honeybee peaks near 10Hz, and falls off rapidly at higher input frequencies, but persists at lower frequencies down to 0.1Hz even for narrow bars. The optomotor response is too slow to stabilise the flight; we must seek an alternative function for it.

Many observations suggest that other control mechanisms exist. Insects flying alongside a stationary wall do not turn in the direction of the relative movement at their side; in fact, they do not turn into the wall

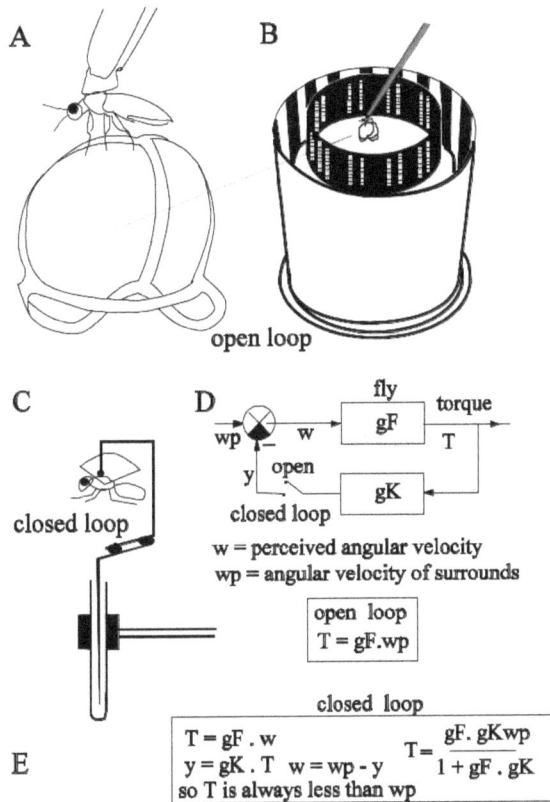

Figure 3. Two ways to demonstrate the optomotor response and analyse interacting inputs and responses. A, A clamped insect walks on an endless strip of paper with choice points at regular intervals. B. The insect at the centre of a drum that has regular stripes on the inside. The inner (black) drum is used to create sequences of modulation that simulate motion. The head cannot move to provide feedback. This arrangement, called "open loop", prevents small spontaneous flicks that inform the visual system about directions of surrounding objects. An alternative is to clamp the head on a rigid pressure detector, and measure the torque exerted by the head (Figure 4). C. An insect mounted on a freely rotating pin so that it can turn freely and detect visual effects of its own motion; called "closed loop". D, E. System interactions; w, motion detected. g, gain. T, the turning motion that may or may not be subtracted from the input.

at all, and they fly readily in a tunnel (see below). Most insects, when standing on a glass plate, turn in the opposite direction to motion of a pattern below them. This response was used by Hecht and Wolf a hundred years ago to measure responses of bees to a moving visual input of controlled intensity, speed and pattern.

In 1940, John Kennedy showed that in free flight, locusts, tzetze flies, and mosquitoes, ignore the optomotor effect, but turn to fly upwind *against* the relative motion of the ground below. Honeybees searching for forage do not look for distant flowers, they fly across the wind using relative motion of the ground, and then turn and fly upwind when they detect an attractive odour. Possibly the optomotor response is used only when an insect hovers and is disturbed by the wind, when all the surroundings are rotated together.

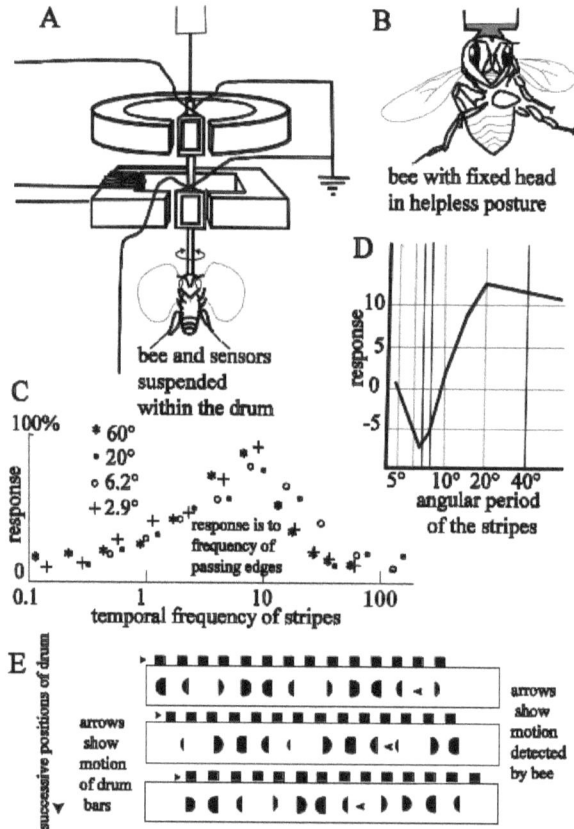

Figure 4. Optomotor responses of the honeybee, and reversal of the response caused by the Moiré effect. A, B. The set-up. C. Responses to temporal frequency (Hz). D. The reversal after the first minimum is reached. E. Three successive positions of the black bars, and the same seen through a line of facets of slight larger separation, causing apparent reversal.

The optomotor response has demonstrated other important points. As in humans, only receptors with spectral peak in the green are involved in all motion perception, and responses to motion adapt to low light intensities. In 1939, Lotte von Gavel, working at Königsberg, found that in *Drosophila* the resolution of black/white bars ranged from a period of 9° minimum in daylight, to 20° as the illumination was reduced. Moreover, the response decreased to zero when the bar width equalled the angle between ommatidia, *then reversed as the bar width was further reduced* (Figure 4D). In bright light, the reversal point was when the bar period exactly covered two adjacent ommatidia, one for the black bar, the other for the white. The reversal is when the periods of stimulus and detector systems come in and out of synchrony (Figure 4E). This called the Moiré effect, easily seen with one window gauze in front of a similar one.

This reversed response shows that detectors of motion and its direction are sensitive to *temporal correlations of modulation in neighbouring receptors, not to the angular velocity or direction of the input over the eye, or the pattern of bars*. Correlation is between neighbouring detectors of modulation (flicker power). This conclusion dominates all considerations of insect motion perception, and must not be forgotten when we analyse landing, avoiding a collision, measurement of range, or recognition of a place.

When the optomotor response, either as the turn of an animal free to move (Figure 3C), or as a torque pushing on a spring (Figure 4A), is plotted against the angular velocity of the drum, the response is at first small then rises to a peak and falls again at higher speeds. In the honeybee, this peak is at about 10 edges passing per second (right side of Figure 4C), which is only about 0.05 of the flicker fusion frequency limit for bee vision. Significantly, the peak is reached at the same frequency of edges passing irrespective of the difference in angular velocity, showing again that the bee detects *modulation frequency* and *not angular velocity*. The optomotor response is too slow to stabilise insect flight, avoid crash, or measure range and speed over the ground.

A bee placed on a flat transparent surface responds to the motion of a regularly pattern of black/white stripes that are moved below by turning to face the direction from which they come, as if to move upwind and preserve a position in space. The resolution of the stripes, measured by Hecht and Wolf (1929) varies with the illumination. At best, in sunlight, the lower limit was a period close to 2°, which is not as good as human vision in the dimmest light. At worst, at 1/10,000 of the brightness, the resolution increased to 20°. Electro-physiology later showed that this adaptation is the same in every ommatidium, not due to a distribution of sensitivity between ommatidia. There must be large reversible internal changes in lateral spacing within the neural processors of motion.

Very early it was noticed that honeybees seem insensitive to the continuous motion caused by movements of the bee itself. As work on peering, scanning, and detection of parallax by various insects has progressed, it has become clear that one function of the optomotor response is *to prevent displacement relative to the general back ground of the distant panorama*, while the nearer edges are detected.

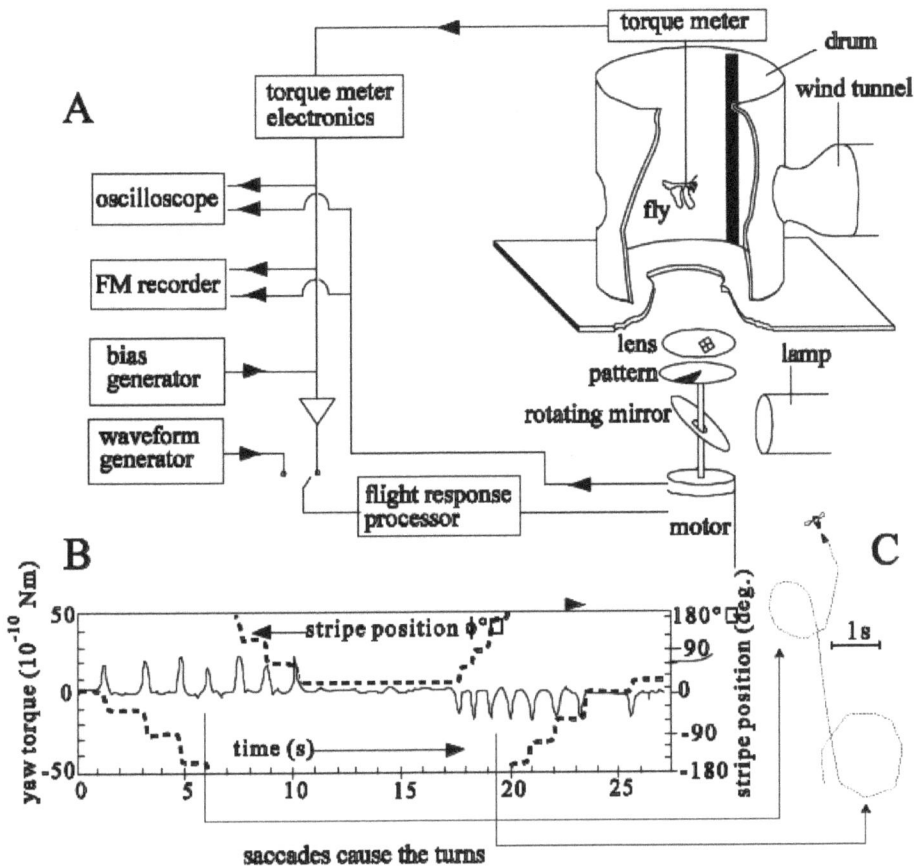

Figure 5. Flies respond to the position of a vertical bar. A. Drosophila in a fixed position, but with visual feedback. The head is fixed but trying to turn towards a vertical black bar that is controlled by the flies own effort. B. For 10 s, the fly tries to make small saccades towards the right (continuous line in the record), and the bar moves in steps to the left (dashed line). After a pause of 7 s, the fixed fly reverses the direction of saccade efforts, and the bar returns. C. The track that the fly in that record would have taken if flying freely.

5. Position of a vertical edge

Common muscid flies, like the housefly, are different. In early experiments on visual control of flight, the fly's head and body were fixed at the centre of the optomotor drum, and the turning force measured (Figure 4), but this yielded nothing but the optomotor response. A long series of experiments by Heisenberg and Wolf had shown that *Drosophila* behaves in sensible and plastic ways when allowed to see *and respond to the consequences of its own effort*. In a white drum with a single vertical back bar in its visual field, *Drosophila* tries to turn towards the bar, in short efforts called saccades (Figure 5). The drum is arranged to respond by making fast movements in the opposite direction, *generating what the insect expects to see as a result of its own effort.* If free to turn, the body would follow more slowly, so that flight would continue in a straight path until the next saccade.

Flies quickly learns how to respond to the unfamiliar situation when the direction of the drum's response is reversed. This explains how it accommodates to wing damage by learning how to adjust the responses of accessory flight muscles continually, so the fly quickly learns to fly in a straight line, although with damaged wings. Similarly, many arthropods rapidly accommodate their gait after the loss of a leg.

6. Modulation is the primary input

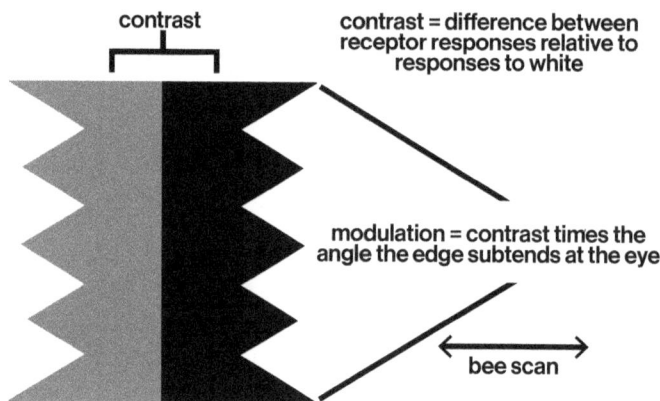

Figure 6. A revised definition of modulation (flicker power), the primary input to the bee visual system. In a scan, the effective stimulus that enters the eye is the length of edge multiplied by the average emission and the contrast at each part of the length. Discrimination tests on bees after training show that they measure and learn total modulation input over the whole eye scan.

In the decade before 2001 we spent much effort to identify the cues that bees use to distinguish and remember the places where they are rewarded. Modulation (see Figure 6) is resolved by single facets. Orientation is resolved by three sets of feature detectors, each three facets long, at 60° to each other in every part of the eye. Bees prefer to learn and remember relative amounts, not absolute values. as shown in tests. Resolution by the known feature detectors was measured, and generalisation of similar signals and ambiguity by the bee is a result of the limited variety of feature detectors. Colour vision remained unsolved at that time.

The first recent breakthrough was the discovery that honeybees distinguish the polarity of any pattern from its mirror image by the left-right relationship of green edges and a patch of blue colour. They also measure the amount of blue in each scan, and blue emission relative to the green of the background. They do not detect black because there is no stimulus, but they measure the horizontal widths between black edges, and amount of blue missing from the white background, because it is displaced by black. They measure *relative total amounts of each*, with minimum differences of about 5%.

Honeybees see the panorama as if they are quantity surveyors on a construction job. When a bee encounters a fence, she sees not the individual planks with green contrast, but a total of the amount of vertical edge and amount of blue relative to green background, but no detail at all. Modulation (contrast times length of edges, Figure 6) is measured in all recognition tasks. Contrast does not vary much in a scene, but the total length of edges crossed in a scan can vary over a huge range, for example with grass or plant stems, so modulation is mainly an indicator of texture. As a result, bees see neither colours nor shapes as we do, but they detect isolated contrasts at edges and appear to be able to count because they measure the total modulation in each scan.

7. Extreme sensitivity to slow and sub-pixel motion

In the mid-1960's, at St Andrews, my own observations of movements of eyestalks of clamped shore crabs in response to motion of a surrounding drum, or pin-light, or celestial motion, showed that they followed the rotation of a drum as slow as one revolution per 24 hours (Horridge, 1966). With nothing else in view on which to stabilize the eyes, crab eyestalks also followed the motion of the sun or moon, and could easily detect the motion of nearby edges of shadows caused by the sun's motion. Is it a clock or a compass?

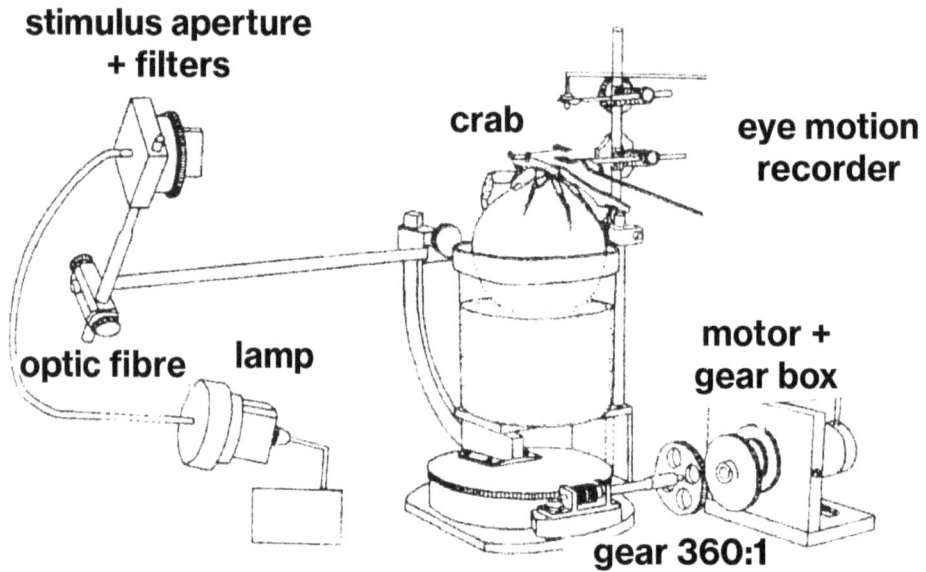

Figure 7. The apparatus that moved a calibrated tiny pin-light at sky speed in front of a ghost crab. (After Doujak, 1985).

One of my students, Fred Doujak, took his apparatus (Figure 7) to northern Australia to test the sensitivity of the large ghost crabs that wander from their burrow over the beach at night in search of food, but are always able to return to them. He moved a tiny pin-light at 15°/h at a measured distance from the eye of a clamped crab, and reduced its intensity until he found the threshold. On his return to Canberra, he took the pin-light and filters to the University Observatory, where the astronomers were able to show that the crab had detected the equivalent of a 1st magnitude star.

Responses to motion of the sun, moon and stars are probably widespread. In 1936 in Holland, Ter Braak showed that some mammals that return to a burrow or cache of their young also have this extra-ordinary perception of optokinetic motion. Dogs were able to detect the direction of motion at 6°/hr, which is comparable to the crab and locust. Dogs and cats are renowned for homing when displaced, and cats stare at the sky. It is all very mysterious, or perhaps not. Possibly there is a widespread sensitivity to direction of sky motion as a universal compass overhead. Responses to extremely slow motion of the sun, moon, or night sky, show subpixel sensitivity when the whole eye is involved.

8. Visual navigation at night

Another former student, Eric Warrant, took a Swedish expedition to Panama, where nocturnal tropical bees navigate home to their brood in the ends of broken hollow reeds, in light levels so low that humans could not see the reeds. Presumably the nocturnal bees had large responses to individual photons, unlike the honeybee, in which the bumps caused by single photons are too small to detect with a microelectrode. From their results it became clear that in bright light the honeybee is 1000 times less sensitive than calculated from the facet aperture, perhaps simply to reduce the energy use and weight of visual transduction processes. If so, our diurnal honeybee is a cheaply built one-purpose insensitive design, solely for use in sunlight, forcing them to go to bed early.

In my laboratory, Eric had worked on dung beetles, which were conveniently available in cultures. They have an eye structure with a thick layer of receptors below a wide transparent zone that sacrifices resolution to get improved sensitivity. In open cattle country in South Africa, these beetles could be studied at work. To keep a straight course while rolling a dung ball under the stars even on moonless nights, a signal of some sort from the sky was essential. Eric concluded that for their compass direction, these beetles use the axis of the milky way across the sky, which would certainly not require acute vision. This adds another possible input for the optomotor response, and earned Eric an Ignoble Prize.

Never resting, Eric hastened back to Australia at the time when bogong moths (*Agrotis infusa*) migrate at night to the high mountains where they aestivate for the summer. There they hang clustered in cool caves, before returning to the pastures where they breed. Captured groups of migrating moths in a large net tent responded to dim illumination from above, suggesting that they could detect a signal from the night sky to guide them in the direction of the mountains.

Many other animals, for example, newts, frogs, and turtles, return at night to isolated ponds, beaches, or river estuaries to breed. Others, such as rabbits, deer, and carnivores return at night to their burrow, or to their hidden place for their young on a grassy plain. Homing pigeons and far-ranging turtles, manta rays and sharks and are still a puzzle. Many birds also migrate over water at night with little to guide them except the sky.

It is very strange that nothing is known about the sensitivity of the honeybee to slow motion of the sky, or to very low light levels. Honeybees use the map of UV polarization (Figure 1), perhaps because they are insensitive to light and must forage in sunlight, but maybe they detect the direction of motion of edges of shadows.

9. Optokinetic memory

In the 1960's, my student Peter Shepheard discovered that after a crab or a locust had a pin-light or a striped pattern in view for a few minutes, the lights could be turned off for a short period, and *when illumination returned, the eye returned to the former position that had been remembered through the dark period.* A shore crab *Carcinus* still had a good memory of its eye's former position relative to a vertical bar *after 10 minutes in the dark.* Locusts remember a shift of 0.1° for 10 s. What could be the use of such a remarkable ability?

When the eyestalk, head or whole animal turns to a vertical bar that had been moved during a short dark period, the optokinetic memory drives the eye back to its original position. With reference to later work, it is important to note that this return, depending on which way it had drifted in the dark, may be *in the opposite direction* to that in the normal response. The former *position, not the motion,* is remembered. Later, Heisenberg demonstrated similar memory of *Drosophila* for the position of a vertical black bar.

At first, we interpreted the memory as a sign that all memory of images was fixed at the stimulus position. It was called "retinotopic" as a feature of the compound eye, but later we realised that the bee remembers the sequence of feature detectors activated in a scan

We called it optokinetic memory, and guessed that it was a side effect of a detector of very slow motion. I have now changed my mind (see below), and consider all optomotor movements to be efforts to keep landmarks at constant positions on the eye, and that the compound eye is perfectly suited to *detection and memory of exact positions, and can also measure and remember the amounts of modulation at each place.*

10. Optic flow; measurement of range

Naïve new born chicks can estimate the range of grain particles against any background without using binocular vision, and peck at them successfully, but only *after their head has moved a bit*. They are able to peck at a rice grain hanging in free space or on a glass sheet, but only after peering with a sideways movement of the head. Similarly, they estimate the height of a small cliff when looking down from the top. In the 19th century, Baron Hermann von Helmholtz described how a moving eye detects relative angular velocity, which is not the same as parallax, and published an exact mathematical representation of the relative motion. He showed that a moving eye could measure the range all around from the different angular velocities in the flow field (Figure 8).

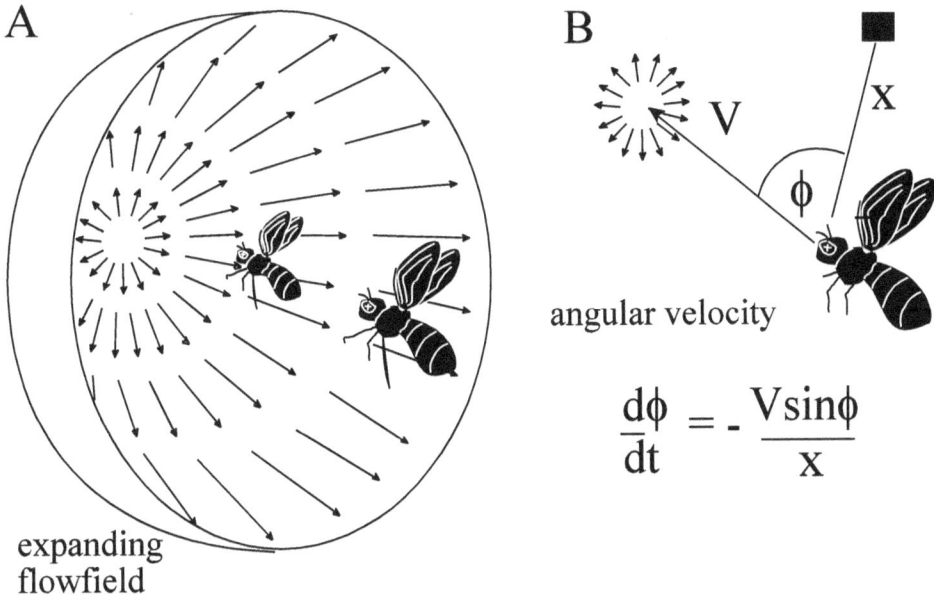

A

expanding
flowfield

B

angular velocity

$$\frac{d\phi}{dt} = -\frac{V\sin\phi}{x}$$

Figure 8. Optic flow. When an eye moves, all surrounding edges appear to move with an angular velocity at the eye, as defined by the equation shown (Figure 8B).

V = forward velocity of insect. x = range of a nearby object.
Φ = bearing of object relative to axis.
dΦ/dt = rate of change of angle = angular velocity in radians/sec

From the carefree way that bees and other insects fly and manoeuvre, they seem to have excellent information about ranges of objects around them. Some predatory insects, like mantis, dragonflies and asilid flies, attack and catch prey in mid-air. The limited literature on vision of fish, amphibia, reptiles and pigeons, revealed similar behaviour; it was all uncanny. Separate objects seemed to be distinguished by relative motion induced by the animal's own motion. Where tested, they could perform against a plain, patterned or even a moving background, or no background at all. However, when striking at a fly, a mantis keeps its head still and estimates the range by binocular collaboration between the two eyes.

The common mantid in Canberra, *Tenodera*, clearly estimated range when stepping over irregularly spaced twigs and leaves. The small sideways movements of the head induce an apparent angular velocity of all surrounding edges that is inversely proportional to their range. When the mantid steps, however, there is no need to peer because the natural sway and forward head motions cause continual relative motion of all surrounding edges. Flying insects also measure ranges of surrounding objects and avoid a crash. They have no need to identify the surrounding objects, only to measure their angular velocity.

In 1987, from a variety of observations on walking insects, I concluded that "perception of an object is inseparable from the local directions of velocity differences. The background may not be perceived at all when an object occurs in the foreground". The system of neurons that detects objects by their motion must retain the directions of those unit motion detectors that are stimulated differently from the motion of the background". Because insect visual neurons detect motion, not stationary contrasts when the eye moves, "the nearest object will then have the highest angular velocity irrespective of whether the eye also has a velocity relative to both object and background". Then later "insects do not have a picture of the world as a spatial pattern of boundaries, edges, contrasts and colours", "they respond as if spatial and temporal discrepancies in the flow field are interpreted as objects". I was convinced that relative angular velocity was the key to insect detection of objects and their spatial relations.

Now consider the honeybee. At that time, it was known that all motion is detected by green sensitive motion detectors that adapt quickly. As a result, a bee with a clamped head learns to give a tongue response to nothing stationary except the colour blue.

About that time, Srinivasan sent me a letter to say that he would like to return to Canberra, so I made great efforts to secure a tenured post. It was probably the last available in our Research School, because times were changing. In Zürich, Srini had found that scanning by bees is usually in a horizontal plane, but they also track up and down vertical edges, and that scanning of edges depends on green receptors alone. He had also found that honeybees were unable to discriminate between a rapidly rotating sectored disc and similar stationary one which also generated a continual flicker from relative motion of the eyes (see Figure 2).

Following a suggestion that we should collaborate on interdisciplinary projects, three Professors, who worked on vision, Bill Levick from Physiology, Alan Snyder from Physics, and myself in Biological Sciences, had persuaded the Vice-Chancellor to found a Centre for Visual Research, This gave us some independence in our own space, with additional funding, a new building, own notepaper, and appointing our technical staff. When Srini arrived, he immediately invited a very skilled bee trainer, Miriam Lehrer, as a visitor.

In a joint venture with Australian Guide Dogs for the Blind, we obtained funds to form a larger group. Srini and Miriam soon showed that motion was also detected by blue contrast. We were joined by Zhang Shao Wu from Academia Sinica, and together quickly showed that somehow optic flow provided flying bees with a third dimension, and they somehow measured range, flight height, and total modulation on a target. It soon seemed obvious that optic flow caused by the bees' own motion was a major part of honeybee vision.

For several years Miriam came to Canberra every southern summer. We tested every idea about vision of spatial relations by trained bees. A useful advance was the use of coloured papers that were equiluminant to one of the types of bee photoreceptors, either green or blue, so that we could separate them experimentally. By March 1988, we had several demonstrations that flying bees detect range of their surroundings by relative motion generated by their own flight (Lehrer, Srinivasan, Zhang and Horridge, 1988).

11. Experiments to test for optic flow

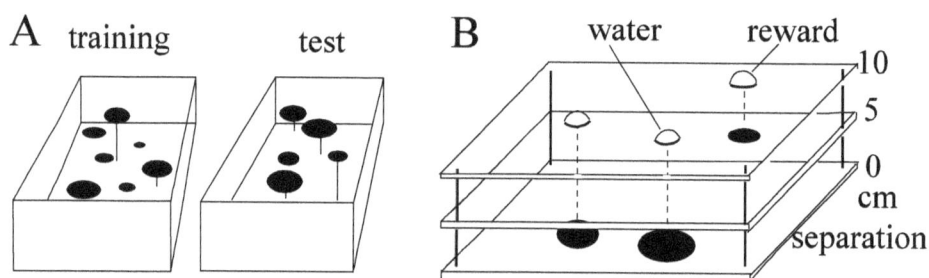

Figure 9. Bees can learn to go the object with the correct height and size among a collection of objects at different heights and shuffled at random A. In a box. B. Between transparent sheets.

A number of discs of different sizes were placed, each on top of a pillar of different height, and their relative positions were shuffled at intervals during training. We trained honeybees to land on one of a particular diameter and height irrespective of location. In some experiments, the whole group of discs was enclosed between glass sheets (Figure 9B).

The bees learned this task, and we assumed that our postulate was validated. However, as I look back years later, I realise that the freely flying bees may have been assisted by a pheromone released by the first bee that by chance found the reward. Also, the experimental set-up was ideal for the use of parallax.

With similar experiments along horizontal flight paths, we showed that flying bees could select the target of the correct range along the horizontal, irrespective of size or relative position (Horridge, Zhang, and Lehrer, 1992). However, I now have similar thoughts that the bees used parallax in the same way as the peering locust, by scanning in flight (Figure 12).

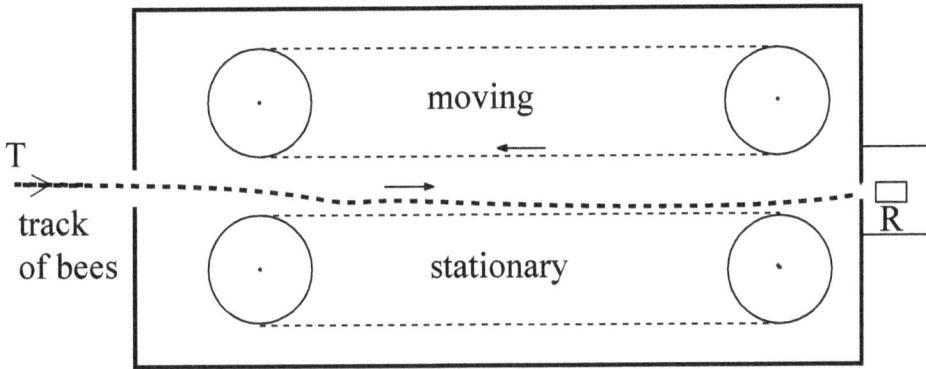

Figure 10. The tunnel between walls that were stationary or moved at a fixed speed in either direction, The bee tracks were photographed from above.

Soon after he arrived, Srini trained bees to fly along a tunnel. The narrower the tunnel, the slower they would fly. In a tunnel with a wall that moved independently on one side, they flew closer to the stationary patterned wall, and moved further away as one wall moved faster (Figure 10). We found that when they had adapted to the situation, bees equalized the apparent angular velocity of each side, even after changes in the patterns on the walls. Srini called it the 'centring response'.

There was an anomaly in that the response was the same for imposed forward or backward motion of the wall. Neuron recording has never found detectors that act in this way; *it must have been related to the avoidance of modulation, not motion*. We had no explanation for that at the time.

Support for target angular velocity as an input came with introduction of an oscilloscope display on one wall of the tunnel, giving greater and faster control of the stimulus at that place on one side. As they came along the tunnel, flying bees found a black/white chequered display at 50Hz or 100Hz, moving at right angles to their line of flight, and their response was observed from the side and from above. As they reached it, the bees moved away sideways. To me, the interesting point is that lateral responses to four different directions of the motion, and to interleaved motions in opposite directions, were all similar, and only double the response to the stationary pattern. There was also a small vertical deviation upwards in response to any direction of motion, and a larger response to a vertical motion upwards. Because this upwards response was only to a low spatial frequency, below 50Hz, it was separately attributed to the optomotor response. The experiment was interpreted as evidence for two motion systems measuring angular velocity, but *we forgot that optomotor responses were to modulation, not velocity*. Looking at this result in 2022, I conclude that the bees saw this additional stimulus as parallax, because they responded to flicker up to 200Hz, as described below in the avoidance response.

With help from the Guide Dogs for the Blind Organization, we obtained funding for additional technical help to make gadgets that detected and calculated the range of nearby objects, initially for visually impaired people. They actually worked. When the camera input was stationary, a screen showed a black/white view. When the box moved, the screen showed the ranges in different directions coded in colour. A miniature eye on a thimble on the middle finger was attached to a output on the wrist, like a Braille output. Despite our efforts, several manufacturers in Australia were unresponsive because few visually impaired people could handle technology (in those days) and few could afford any expense.

Anyway, after the Chernobyl nuclear power station exploded in April 1986, the Russians sent in men who soon died. Having 50 nuclear reactors and no alternative, the Japanese sent out engineers to search the world for visual systems for mobile robots, and a group arriving in Canberra found my project, as told in my book. The Fujitsu Computer Company invited four professors to visit their factory, but I said nothing there about relative motion. Eventually they gave ANU a sum of A$10 million for our ideas, so we expanded and applied our system to the control of freely moving vehicles and drones with a computer on board.

12. Measurement of distance flown

In the decades before 1940, von Frisch had concluded that the distance flown on the foragers' homeward flight with their food was measured, and on arrival conveyed to other bees by the number of waggles in their dance. Because they recorded a greater distance after flying uphill, he concluded that they measured the amount of sugar used as fuel. One of his students even measured the weight of sugar used by marked bees on a flight, and "proved" his idea.

In a novel experiment in 1995, Esch and Burns counted the waggles in the dance of bees returning from a feeder located on a distant balloon that had been raised to various heights over a constant terrain. The number of waggles decreased as the balloon was raised, which suggested that their bees had measured the distance flown *by integration of the apparent angular velocity of objects below*. However, at a greater height, they would also detect less modulation, but this was overlooked.

In 1991 Mike Ibbotson had published physiological support from his earlier work. In a survey of 65 descending neurons in the ventral nerve cord of the bee, stimulated by bars moving on a screen, he found most of them responded to modulation, but 12 that apparently measured angular velocity in the range of normal flight, irrespective of spatial or temporal frequency. This conclusion stuck, and influenced our subsequent research, but the function or connections of these neurons is still unknown.

About the same time, Srinivasan and his group demonstrated that bees flying into a tunnel, with a pattern of contrasts *that does not change* on the insides of the walls, measured the distance flown from the entrance (Figure 11). They assumed that their bees measured distance flown *by integration of angular velocity*, but modulation would serve equally well. Later it was shown that bees flying uphill flew closer to the ground which gave them faster visual feedback from the ground, but also the lower altitude would be accompanied by an increase in modulation below.

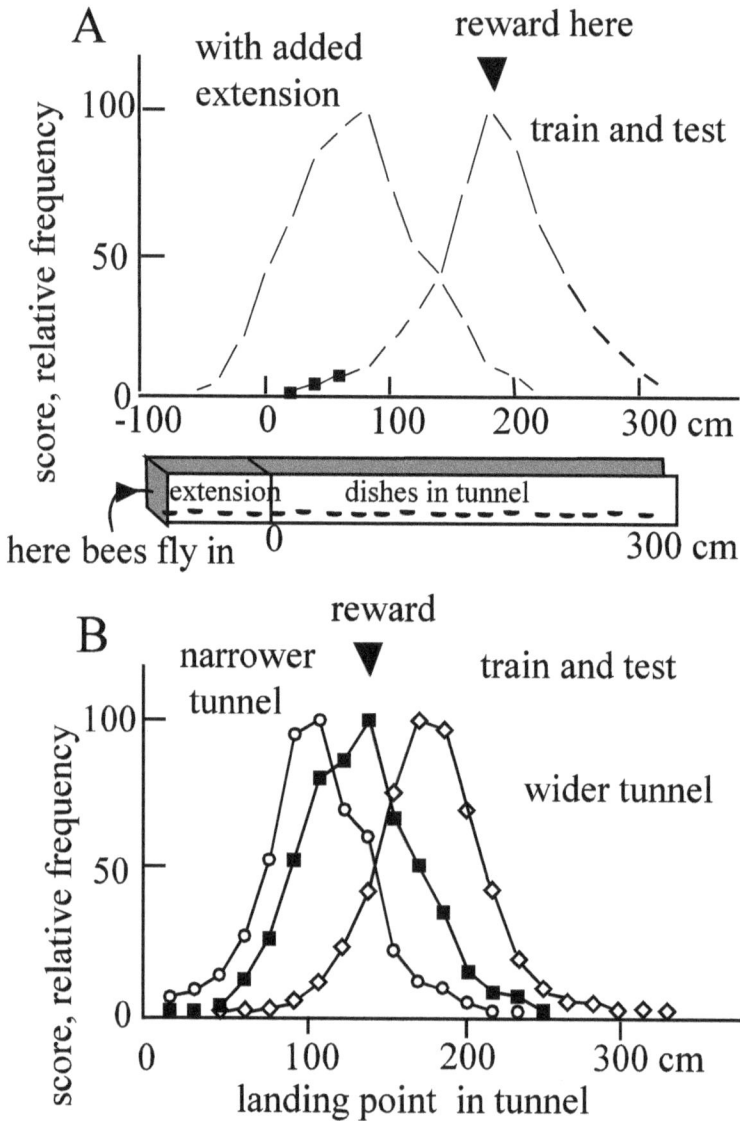

Figure 11.A. Trained bees entering the tunnel flew for a particular distance then selected a dish where they expected to find a reward. The result with the extension shows that they learned to measure the extra distance from the entrance of the extension. B. In a narrower tunnel the trained bees went a shorter distance. In a wider tunnel, they went further. Whether they detected modulation was not tested.

Much later, in 2004, unfortunately Srinivasan and his group showed that bees flying over water *innately* made less waggles in the subsequent dance, and he also demonstrated that bees detect *a component from the total amount of edges passed*, which depended on the amount of detail along the route. The homing bee takes a measure of total modulation in the visual experience in flight (contrast intensity). These careful observations *of innate behaviour* by multiple observers confirmed my doubts whether bees measured angular velocity at all.

Long ago, Laughlin and Hardie (1978) showed that the motion-detector neuron pathway of insects is optimized to detect motion of edges. At the time we were unaware that bees preferred to learn relative features rather than absolute measures, and distinguish different sizes by amount of blue emission, or by width between vertical edges of green modulation. We had ignored the continual scanning by bees, parallax, the optomotor memory response to a change in place, and the sensitivity to the position of a vertical edge. Instead, we continued to assume that they learned to discriminate the correct range and size of the rewarded target from the angular velocities induced by their own movements. Earlier superficial examples of tracking or chasing, by flies, dragonflies and mantids, had led to the same conclusion. Also, we ignored some anomalies, always a fatal thing to do.

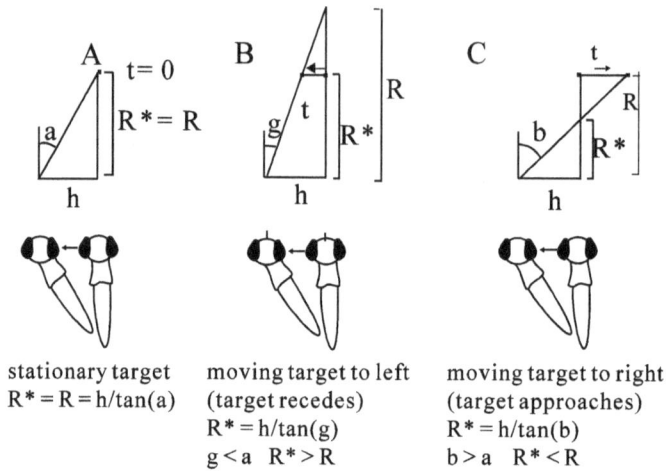

Figure 12. A. A locust looking at a stationary target makes a sideways movement h, without rotation of the head, and then jumps a distance R*, which equals the correct distance R, B. Motion of the target, t, in the same direction as the head causes the jump to overshoot the target with the movement R > R*. C. Motion of the target in the opposite direction to the head causes the jump to overshoot the target with the movement R* < R.

13. Peering

Locusts measure range before a jump by making a small sideways movement of the head, called peering (Figure 12). Common mantids that I observed clearly estimated range when stepping over irregularly spaced twigs and leaves. They measure the range of nearby edges all the time, as they move. When a mantid steps over twigs, there is no need to peer because the natural sway and forward head motions cause continual apparent motion of all surrounding edges. It is ambiguous whether the jump distance depends on the optic flow (Figure 5) or the change in position (Figure 7), or parallax, see below.

14. Parallax

Flying insects also measure ranges of surrounding objects, and avoid a collision. To pilot themselves, they have no need to identify surrounding objects, only to measure their relative changes in position. For a foraging bee, this feedback might also serve to identify different places.

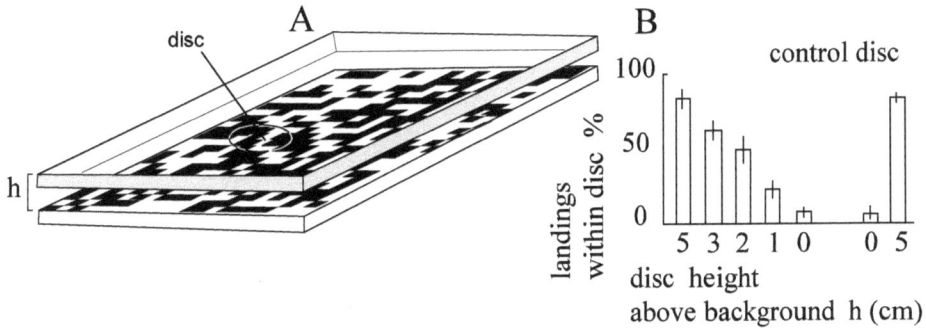

Figure 13. A. Bees can easily locate a small disc that is raised above a background of the same random pattern as the disc, because parallax is generated by the motion of the bee itself. B. If rewarded, the percentage landing on the disc is proportional to its height. (Redrawn from Srinivasan et al., 1990).

15. Innate detection of parallax

The innate detection of a checkered disc above a similar check background was demonstrated in 1993 by Miriam Lehrer. As the disc was lifted above the background, recognition improved (Figure 13B). At a feeding place where the reward was in the centre of a piece of white card raised above a patterned background, *90% of the bees landed facing inwards* at the nearest edge of the card where they detected closing parallax. On the other hand, when the reward was on a transparent glass plate surrounded by background, with another background visible below, *90% of the bees flew over the glass to the far edge and landed facing outwards*, again on closing parallax. In each situation, they would not land on opening parallax (defined in Figure 14).

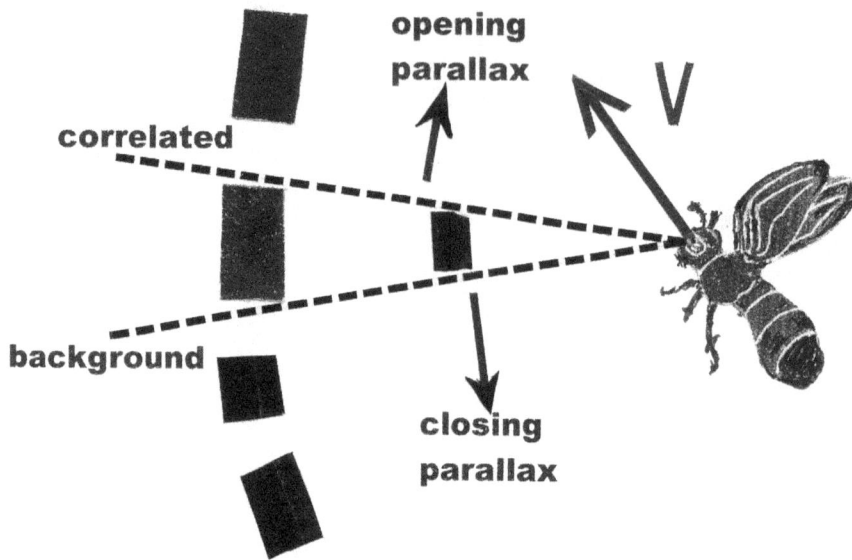

Figure 14. Opening and closing parallax.

Miriam Lehrer noticed that freely flying bees preferred to land at right angles to edges or boundaries, but only when green contrast was present. We analysed this situation and discovered much more (Lehrer, Srinivasan and Zhang, 1990). "A randomly structured pattern cannot be detected against a randomly structured background, and cannot be detected by the bees unless it is raised above the background." "In detecting the figure, the only cue appears to be the *apparent motion of the nearest edges relative to background*". Only the relative motion was considered. We called it "boundary parallax", described various varieties with edges oriented in different directions, and gave an explanation with a model which looks at the difference between neighbouring motion detectors.

Bees spontaneously locate an edge where one surface is raised a little above the other (Srinivasan, Lehrer and Horridge, 1993). They always approached a little cliff at right angles to the edge, and landed only when they approached from the lower surface. Changes of patterns on each side of the edge had little effect on the results. Forgetting about parallax, it was concluded that they landed "*by the local change from slow to faster motion* that is experienced when the bee approaches the edge from the lower side (Lehrer and Srinivasan 1993).

The next step is very significant. As they fly around over an unfamiliar patterned surface, at first in random directions, bees discover places where two different levels show parallax. Then *they change direction and fly at right angles to the edge*, which maximizes the parallax (Lehrer and Srinivasan, 1994). This is quite voluntary on the part of the bees, and shows that they have an intrinsic mechanism that persists without training or habituation, and enables them to extract three-dimensional structure in flight, without being in an artificial apparatus where they learn to make the experiment succeed in an un-natural situation (Figures 8, 9).

Usually, honeybees scan gently from side to side as they forage. Scanning is an excellent way to estimate the range of objects in view, because they can then distinguish the nearer edges from the more distant background by the parallax that they generate.

16. The avoidance response

Before he returned to Canberra, Srini had been working with Professor Rüdiger Wehner in Zürich on the temporal resolution of the flying bee towards a rapidly flickering rotating disc, diameter 10cm, that offered a reward (Figure 15). It was called the 'avoidance response' because the bees would not come close to the flicker, although they were hungry and trained at 6Hz to go there. The avoidance was maximum at 60-90Hz, and cut off above 200Hz, at about the same limit as the retinal receptors. This contrasted with optomotor responses of the bee that were maximum at 2-3Hz and cut off at 20Hz (Srinivasan and Lehrer, 1984).

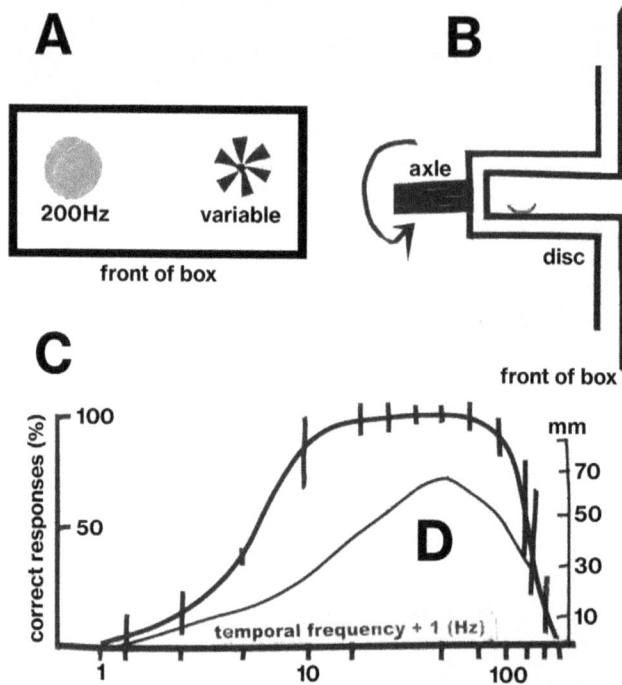

Figure 15. The avoidance response. A. The front of the box as seen by the bee, with a sector disc on the right, and a similar disc moving at 200Hz, both seen through the transparent front wall of the box. B. Seen from the side, the discs rotate at a controlled speed, and a reward for the bees lies within the cavity on the outside of the wall. C. A plot of the success rate (left hand scale) when bees trained to prefer the 200Hz disc had to choose between it and the variable disc at different flicker frequencies. D. The distance from the target (right hand scale) where bees stopped and would go no further to the target, although trained at 6Hz to go there. (redrawn from Srinivasan and Lehrer, 1984).

Bees were unable to distinguish a stationary disc from the same disc generating 200Hz of flicker, exactly as when tested with flicker without motion (Figure 2), because they responded to the total flux, which was not altered by the motion of the disc. However, they were easily trained to distinguish the discs when the variable disc generated flicker at 6 to 10Hz, and in this range they could detect the radial pattern..

With flicker above 10Hz, although *the bees could choose the correct disc, they would not go too close*, and stopped at a limiting distance up to 70mm from the reward (response curve D in Figure 15). This was astonishing. They had been rewarded there in the training, but now found a barrier in the air while in free flight. The halt in the flight was an intrinsic response to something that they all easily detected.

My present opinion is that bees fly among foliage or close together, so must be able to detect and avoid obstacles and the wings of bees beating nearby. What did they see in this experiment? Certainly, it was not only the flicker.

What they saw was strong illusion of *meaningless parallax generated by the different motions of the edges of the sectors of the variable disc*, in the frequency range which indicated that they were very close and might collide, with no meaningful parallax that indicated a nearby edge suitable for landing. They wanted that reward, but there were *too many edges lashing about in meaningless directions* to allow them to settle. Finally, at 200Hz, when the correct choice fell to 50%, the bees could not distinguish the moving disc from the stationary because they caused an equal response to each target (by Talbot's Law, as in Figure 2).

17. A place for the optomotor response

In an animal with relatively little brain that needs to be lightweight to fly, the use of parallax to sense the three-dimensional world introduces an ambiguity. When you focus on a nearby object and move your head, you notice that distant objects appear to move in the same direction as your head. When you focus on distant objects, the change in position of nearby objects is magnified, and they appear to move in the opposite direction as your head. The input of the optomotor response is always from the whole visual field of both eyes. The average of the background positions provides a maximum stability of the head and eyes is always available, and the system as a whole can identify the nearest object, and prevent a crash.

The optomotor response, combined with the optomotor memory of the first position of an edge, and preference for closing parallax, appears to be sufficient for control of flight. This brings together all the *innate* responses that have previously been studied separately. The next step is to devise a suitable stimulus so that we can search the insect brain for neurons that respond in the expected way. Motion detector neurons usually have very large fields but those that detect range from closing parallax must have two opposed small directional fields and will probably be small local neurons. This makes the electrophysiology rather difficult.

18. We had the wrong model

Right from the start, we had the wrong explanation of 3-D vision of the bee. When astronomers calculate the 3D arrangement and distances of heavenly bodies, they do not use relative velocities, they measure the relative positions of each object on photographs taken at different known times.

Now, hold up your finger, look at it, and move your head. *The most obvious change is a shift in position of your finger; you never noticed its angular speed.* When you drive in traffic, do you measure the angular velocities of the other cars? I believe that when Helmholtz walked among trees, he actually saw their arrangement as a wood, by changes in their positions, not their angular velocities.

My doubts were also based on the nature of inference, and how it differs from deduction and leads to erroneous conclusions when experiments are designed to produce the expected result. Experiments on relative range with the discs on pillars or objects at different distances (Figures 8, 9), were planned to test whether bees detected differences in angular velocity, but that is a hard variable to calculate, requiring a measure of time elapsed in the action, whereas looking at successive positions does not involve time elapsed. Just because our experiments worked, does not prove the postulate about angular velocity. The bees had an opportunity to learn to make the experiments work. On the other hand, many direct observations show that *naïve bees detect relative differences in position; in parallax, in optomotor memory, and in navigation by landmarks*, and can also learn the width between two vertical edges.

All the experimental tests whether honey bees could measure range from optic flow were of performance only; i.e., whether they could do the task. However, to discover what the bees actually detected with their eyes, it is essential to make many tests of trained bees to reveal the actual feature detectors that they used.

Much later, when I trained flying bees with a variety of targets and with the avoidance response, *and tested them to discover what they actually detected*, I found that they always measured and learned modulation, which is the length of edge detected, multiplied by the contrast in each bit of edge (Figure 6). They also detect and learn left-right asymmetry of modulation, the position of a vertical green contrast and its position with respect to a patch of blue, and also the total blue content at each scan, all measured as a running total of modulation in a scan. When there is no green modulation, they can detect, measure and learn total blue modulation. Modulation and its position is what they detect and learn, and what accounts for these responses; not angular velocity.

There was another persistent problem with use of angular velocity to measure range. We often wondered why the motion detection of all insects we studied had the maximum possible resolution, as limited by wavelength, aperture and receptor width, and also adjust the resolution to be the maximum at reduced light intensities. After all, a kind of measure of angular velocity can be obtained by any leaky integrator, apparent angular velocity is different in each part of the eye, resolution means less for moving objects that may be blurred at later stages of processing, and a measure of velocity requires a measure of time and two positions.

If angular velocity was the way to measure range and distance flown, honeybees would not need to use the less exact measure of modulation, which is not constant from place to place, but measurement of parallax requires maximum resolution of edges.

19. Our application used another model

All this work was ignored when we started to put insect vision into robot aeroplanes. I had started the project with a solid belief that the insect mechanism was designed to measure ranges of surrounding objects from the optic flow, using equations of relative motion done so well by Helmholtz (Figure 8). The mathematicians loved it.

However, when our circuit designers first constructed the eye of their flying machine, they were limited to an input of a camera raster of pixels stuck in rows on a screen. Calculation of the distribution of angular velocities from successive intensities in pixels generated by a moving camera is an extremely tedious process, even for a small area of the visual field, because a correspondence of each pixel with its neighbours must be matched with the same place in the next camera frame. This is why we use the human brain to fuse the successive images when watching television.

This problem was *avoided* when the camera looked at the scene through *two adjacent mirrors at a small angle to each other*. Statistical correlation between a patch in one mirror with that in the other mirror gave a measure of the range. In fact, the measure was the mismatch derived from the parallax created by the angle between the mirrors.

About 2008, Srinivasan moved to Brisbane and our research group in Biological Sciences was all moved to Defence Research in Adelaide to continue the work in secret. On April 26, 2021, a drone helicopter flying on Mars called Ingenuity, was obliged to travel slowly when mapping surface features, because limit*ed by its visual processing system*. Now, a new fighter drone, The Loyal Wingman, from Adelaide, is presumably the latest development in conjunction with Boeing, but apart from radar, I have no idea how its senses its three-dimensional surroundings. Nowadays, handy small lasers are used to measure range, and the many sensors now available are nothing like those of a bee.

Having digested all that, dear reader, you may understand that animal visual systems have no raster or pixels. They function by hard-wired detection of specifiable visual features to which they have become adapted by long and effective natural selection. They detect features of value to them, that we discover only by training followed by exhaustive testing. A robot explorer is different. It cannot see the pictures taken by its own cameras or by the drone flying above, but only pass back the images to be interpreted by a human brain. Maybe, with a large number of suitable feature detectors, we will make self-drive vehicles, that see effectively in real time, and respond in an adaptive way.

You may ask what is the relevance, therefore, of further understanding of insect piloting. All animals and man use parallax to reveal the 3D world. Robots, drones, and self-driven cars need the same information. We should make use of mechanisms already selected and successful in the natural world. However, development of self-driving cars has been ineffective and slow, and if we are to develop a lightweight electric flying car, a pair of head lights and a rear-view mirror will not be sufficient. Parallax is in the instruction manual for 3D piloting, and we would be stupid to ignore the well-adapted animal visual mechanisms that provide new ideas far into the future.

References

Doujak, F. E. Can a Shore Crab See a Star? J. Exp. Biol. (1985) 116 (1): 385–393.

Horridge, Adrian. 2021 Honeybees Vision: Recent Discoveries. Published by www.northernbeebooks.co.uk

Horridge, G. A. 1966 Direct response of the crab *Carcinus* to the movement of the sun. J. Exp. Biol. 44, 275-283.

Horridge, G. A. 1986 A theory of insect vision: velocity parallax. Proc. R. Soc. Lond. B 229, 13-27.

Horridge, G. A. 1987 The evolution of visual processing and the construction of seeing systems. Proc. R. Soc. Lond. B 220, 279-292.

Horridge, G. A. and Marcelja, L. 1992 On the existence of fast and slow directionally-sensitive motion detector neurons in insects. Proc. Roy. Soc. Lond. B 248, 47-54.

Labhart, T. 1980 Specialized photoreceptors at the dorsal rim of the honeybee's compound eye; polarization and angular sensitivity. J. Comp. Physiol. A 141, 19-30.

Lehrer M, Srinivasan M V, Zhang S W and Horridge, G. A. 1988 Motion cues provide the bees' visual world with a third dimension. Nature, 332:356-357.

Srinivasan M V, and Lehrer M, 1984 Temporal acuity of honeybee vision: behavioural studies using moving stimuli. J. Comp. Physiol. 155, 297-312.

Srinivasan M V, Lehrer M, Zhang S W and Horridge, G. A, 1989 How honeybees measure their distance from objects of unknown size. J. Comp. Physiol. A 165, 605- 613.

Srinivasan, M V, Lehrer, M. and Horridge, G. A. 1990 Visual figure-ground discrimination in the honeybee: The role of motion parallax at boundaries. Proc. R. Soc. Lond. B. 238, 331-350.

Wolf, E. and Zerrahn-Wolf, G. 1935. The effect of light intensity, area and flicker frequency on the visual reactions of the honeybee. J, gen. Physiol. 18, 863-864. and 1935. The validity of Talbot's Law for the eye of the honeybee. ibid. 865-868.

For further references, see my book:

Horridge, Adrian. The Discovery of a Visual System: The Honeybee. CABI Books. Boston, USA and Wallingford UK 2019